BURT FRANKLIN: BIBLIOGRAPHY & REFERENCE SERIES 411

THE LITERATURE OF PAPERMAKING 1390–1800

The Mountain House overlooking Chillicothe
from a print of 1852
Dard Hunter's Private Press

Dedicated
to the Memory of
My Mother
Harriett Rosemond Hunter,
1858-1925.

THE LITERATURE OF PAPERMAKING 1390-1800

BY

DARD HUNTER

BURT FRANKLIN
NEW YORK

Published by LENOX HILL Pub. & Dist. Co. (Burt Franklin)
235 East 44th St., New York, N.Y. 10017
Originally Published: 1925
Reprinted: 1971
Printed in the U.S.A.

S.B.N.: 8337-17693
Library of Congress Card Catalog No.: 68-56797
Burt Franklin: Bibliography and Reference Series 411

Reprinted from the original edition in the Ohio State University Library.

THE LITERATURE OF PAPERMAKING 1390-1800

PREFATORY NOTE

DURING the fifteen or twenty years in which I have been interested in handmade paper, and also in the history of papermaking, a large number of books relative to these subjects have been collected. It has been my wish in this bibliography to form a permanent record of the works which had appeared before the year 1800, as no other attempt has been made along this line of research.

A select collection of material upon a subject has always seemed to me to be a necessary preliminary step to its study, and this compilation is published in hope of encouraging research in the interesting art of the early craftsmen who fashioned paper by hand, and made it possible for the "art preservative of all arts" to hand down to us the literary and graphic art treasures which have accrued since the invention of printing.

It will be noted that the books and pamphlets listed are all by European or American writers, as I have made no effort to include any item of Oriental origin, nor has there been any desire to enumerate the many encyclopedic articles of any language.

With one or two exceptions I either possess copies of the volumes listed or have examined the works in libraries. I have set most of the titles and descriptions of the books directly from the type-case without notes or copy, but with the original volumes and pamphlets before me.

All the title-pages have been reproduced directly from the old books and have been printed upon paper as nearly like the original as possible so that each facsimile might resemble the antique title-page in every way. The photograph of the two pages from the Ulman Stromer manuscript was kindly furnished by the director of the Germanisches National Museum in Nürnberg, where this interesting document has been deposited for years.

Dard Hunter,
February 7, 1925.

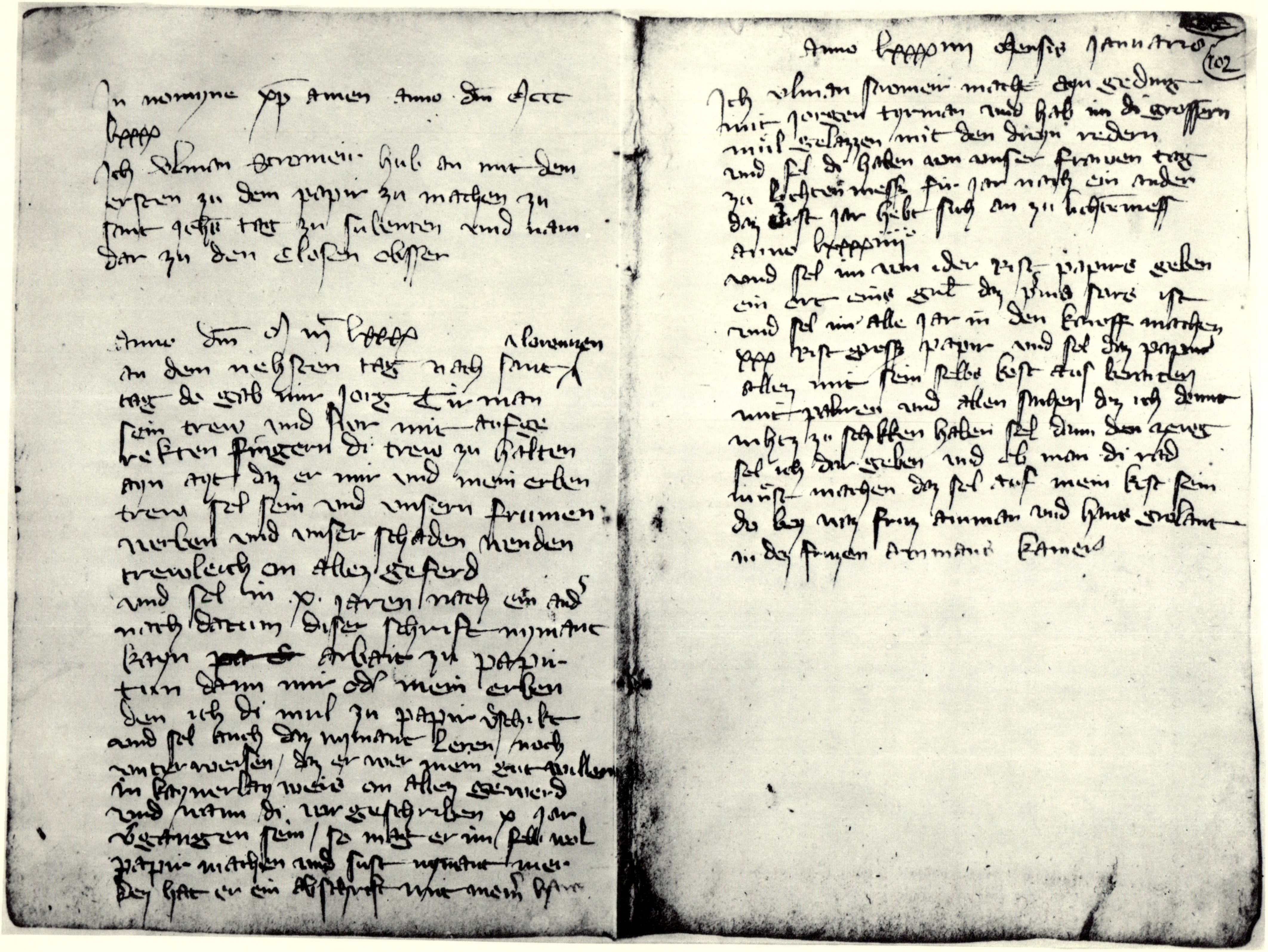

Two pages from the manuscript written by Ulman Stromer in 1390. From the original in the Germanisches National Museum in Nürnberg.

THE LITERATURE OF PAPERMAKING 1390-1800.

STROMER, ULMAN, 1328-1407.

Püchl von meim geslecht und von abentewr. (Manuscript in the hand of the author.) Nürnberg, 1390. 15 x 21 cm.

The first writer to touch on the art of papermaking was Ulman Stromer who compiled a diary which dealt with his tribulations in founding a paper mill. Stromer was a member of the senate that governed Nürnberg and his mill was located just outside the western wall of the city at the point where the Pegnitz River emerged into the fields. Before the papermaking venture Stromer had been a merchant and on his trading journeys to Northern Italy he had seen paper mills in operation and was convinced that he would be successful with an establishment of the same kind in Germany, especially if able to secure a monopoly in it. No one in Nürnberg knew anything about making paper so it was necessary for the erstwhile trader to procure his employees elsewhere. He set out to find several foreigners who were skilled in the art of papermaking and after considerable deliberation Stromer was successful in inducing a number of trained workers to leave their positions in Lombardy.

Ulman Stromer at this time was sixty-two years of age and his experience in trading had made him keen and of a cautious disposition. Therefore, he had all of his employees swear to be loyal to him and to keep the secrets of their craft to themselves so that no other paper mills might be springing up in Germany.

Accordingly Stromer wrote in his diary that Closen Obsser, the man he had engaged as foreman, "promised to be faithful unto me and declared on his oath that he would be true to me and my heirs, that he would be my overseer at the mill, keep me from harm and that as long as he lived he would make paper for no one else save me and for my heirs and would not teach any man to make paper in any way at all. This took place on the Sunday next before St. Lawrence Day, in my room at the time of evening prayer in the year 1390 when my son Jörg was present." Thus having the foreman's faithfulness assured Stromer proceeded to have the rest of his employees take the same oath. He recorded in his chronicle that, "it was on the day after St. Lawrence when Jörg Tyrmann swore on oath to the Saints that he would help my progress materially and that for ten years he would make paper only for me and my heirs and would not instruct anyone except at my sanction. But when the ten years are passed he may make paper for himself but for no one else and he may then teach those who make paper for him, but for no one else, as long as he lives."

These men, like Stromer, were Germans but with the skilled Italian artisans he felt an extra precaution essential so he took them before a procurator named Conradus and Stromer's sons, brother, and brother-in-law acted as witnesses to the affidavits. The contract with the Italians read thus: "In the year 1390 Franciscus de Marchia, and Marcus his brother and his manservant Bartholomeus pledged their loyalty to me and swore on oath to the Holy Saints that they would forever be faithful and would not divulge the secrets of papermaking to anyone in all the German lands this side of the mountains of Normandy."

The Stromer mill from an engraving in Schedel's "Liber Chronicarum," printed by Anthony Koburger, in Nürnberg in the year 1493.

The precautions taken by Stromer to guard his own interests rather excited the suspicions of the Italians. They felt if they were so indispensable to the success of the paper mill that it would be better for them to control the enterprise than for Stromer to be the master. With this idea in view the Lombardians tried to hamper the paper mill on every side hoping Stromer would become weary of the whole venture and lease the mill to them at a low figure for them to do with as they liked. Continuing with his narrative Stromer wrote, "the Italians were most troublesome the first year and did as they pleased and caused me many difficulties and would not suffer the third wheel to run at all. My first two wheels ran eighteen stampers and even these were left idle a great part of the time as the foreign workmen wished to produce as little paper as possible so as to force me to let them have the mill for a rent of 200 gulden a year. This I would not agree to do and they then offered also to give me an amount of paper. From this I understood that they wished to ruin me and deprive me of the paper mill."

In medieval times the master had great authority over his workers and Stromer recorded in his diary: "In the year 1391 on the twentieth of August, I took Franz and his brother Marcus and shut them in the tower." On the fourth day of their imprisonment the unruly Italians sent for Stromer's brother-in-law, thinking that he would be more lenient than their master. A reconciliation was effected and the men were liberated on the fourth day but not until they had sworn that they would cause no more trouble and that they would thereafter do their full share of work at the paper mill.

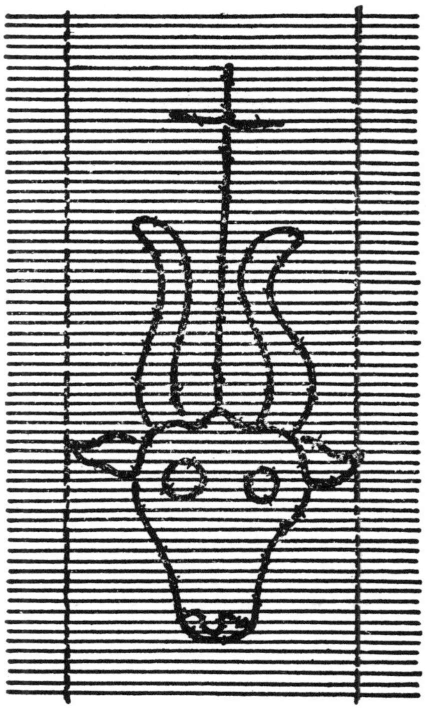
The watermark of the bull's head.

The confinement in the tower had the desired result upon the scheming Italians and they worked more in harmony thereafter. Encouraged by the improved discipline Stromer engaged additional papermakers who were duly sworn like the others. A carpenter was needed to repair the stampers and vats, for in 1392 Stromer recorded: "In my back room Erhart Zymerman has been engaged to be my servant for one year. He is to do the carpentry work at the mill or he will polish paper, and his wife will sort rags, or hang paper on lines to dry, or count the paper. Wages will be paid them, a good house in which to live and firewood withal."

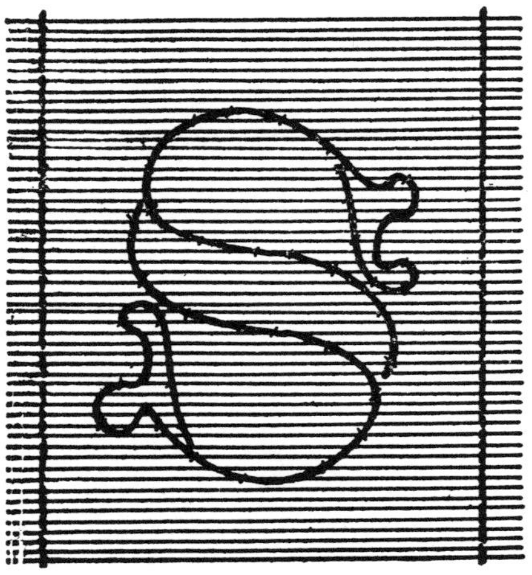
The "S" watermark used by Stromer.

Stromer operated the paper mill from 1390 to 1394 when he leased the property to Jörg Tyrmann for a period of four years.

The paper on which the Stromer manuscript was written reveals the bull's head and cross watermark in three different forms but it is thought that the paper was not from the Stromer mill but of Italian fabrication. The chain-lines of the paper are widely spaced and the laid-lines heavy and coarse. This volume which constitutes the most valuable and unique document in the annals of papermaking is now in the German National Museum in Nürnberg.

An initial "S" was used on some of the Stromer moulds as a watermark and also the arms of Nürnberg was used as a device in part of the paper that was fabricated in this fourteenth century mill. Stromer died in the year 1407 at the age of seventy-nine.

RIPOLI PRESS.

An original Cost-book which was compiled by the directors of the Ripoli Press. Florence, 1474-1483. 13 x 19 cm.

This manuscript volume is now in the Magliabechian Library, Florence. The book is interesting as it gives the prices of paper and printing materials in Italy in the fifteenth century. Nine sizes of paper are listed as being in use at that time. The names of these papers include Bologna, Fabriano, Colle, Prato, and Pescia.

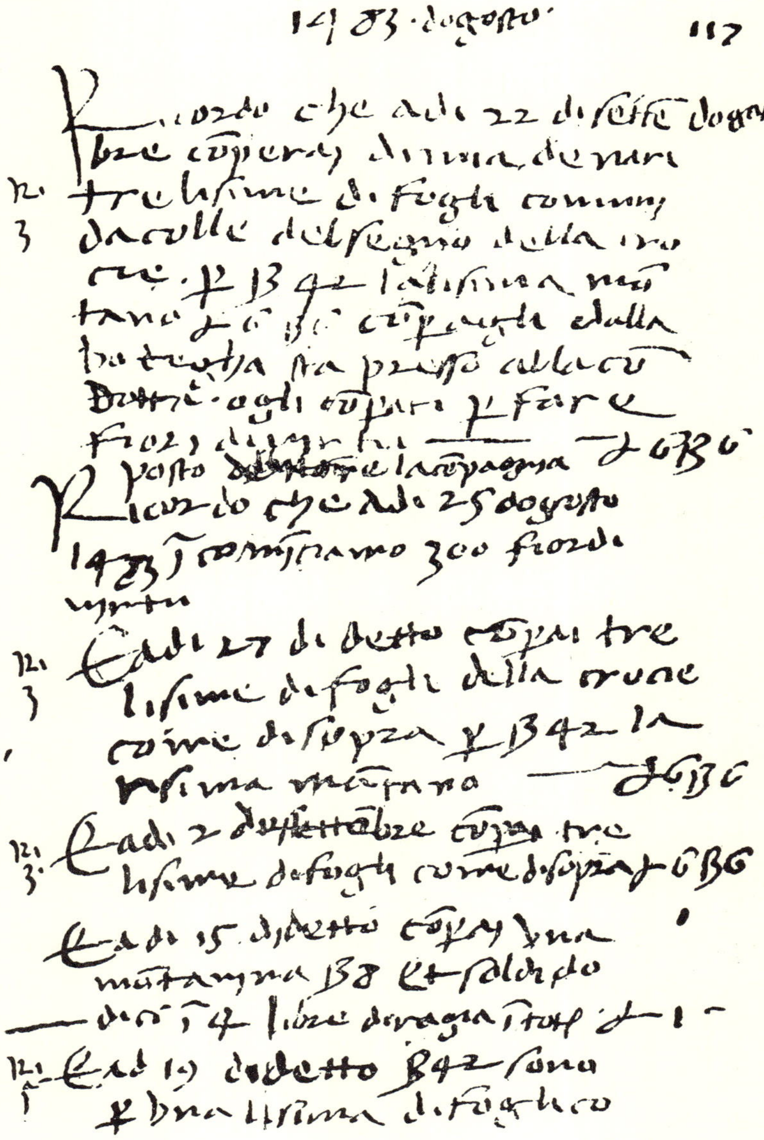

A page from the Ripoli manuscript.

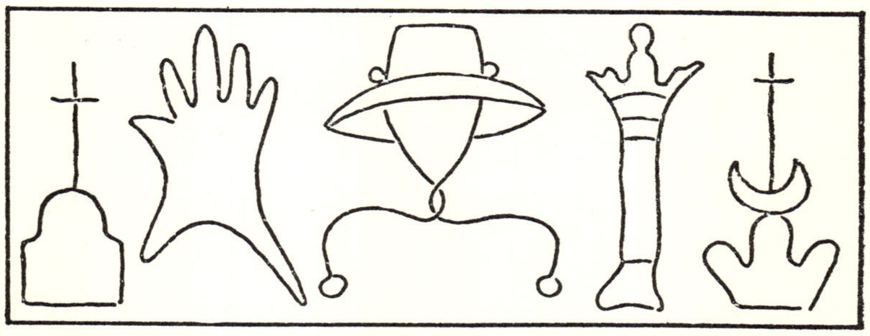

The five watermarks found in the paper upon which the Ripoli manuscript was written.

BARTHOLOMAEUS: DE PROPRIETATIBUS RERUM.

Printed by Wynken de Worde, the successor of William Caxton.
Westminster, 1498.

This volume is supposed to be the first book printed in England in which English paper was used. The paper was made by John Tate, who operated a mill in Hertford. The watermark used by Tate was in the form of an eight-pointed star enclosed in a double circle.

John Tate was a mercer in London, and the son of John Tate, who was mayor in 1473. He died in 1507, and his will contained several references to the paper mill. In the first place he bequeathed to Thomas Bolls of Hertford, "as moche whit paper or other paper as shal extende to the somme of xxvi s. viij d. . . owte of my paper myll at Hertford." Then he directed his executors to dispose of the paper mill "with all the goods, woods, pastures, medes, with all the commodities concerning the said myll to the moost advantage." Finally, in leaving to his eldest son Robert all his lands in Hertfordshire and Essex, the paper mill was excluded: "My paper myll with the appurtenaunces to be alway excepted and to be solde."

The Tate mill was evidently short lived as there is no evidence to show of its existence before 1495 or after 1498.

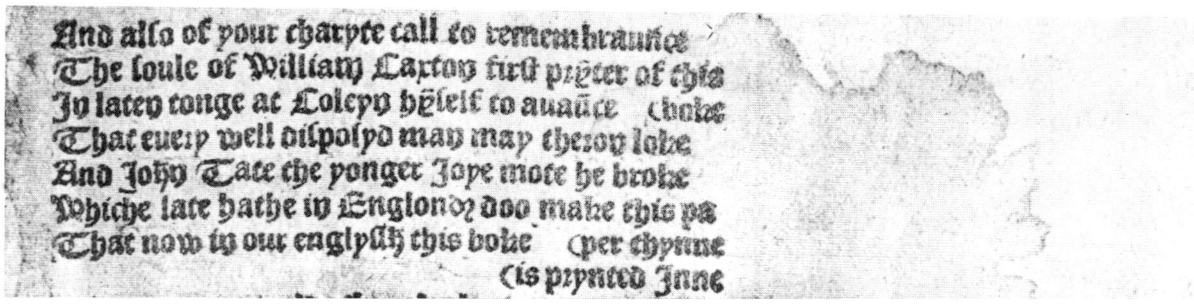

And also of your charyte call to remembraunce
The soule of William Caxton first prynter of this
In laten tonge at Coleyn hymself to auaunce (boke
That euery well disposyd man may theron loke
And John Tate the yonger Joye mote he broke
Whiche late hathe in Englond doo make this pa
That now in our englysshe this boke (per thynne
(is prynted Inne

The verse in De Proprietatibus Rerum regarding the paper upon which the volume was printed.

SACHS, HANS, 1494-1576.

Eygentliche Beschreibung Aller Stände auff Erden, Hoher und Nidriger, Geistlicher und Weltlicher, Aller Künsten, Handwercken, etc.
Francksfurt am Mayn, 1568. 15 x 21 cm.

The earliest engraving depicting a papermaker at work may be found in this book of trades. The wood engraving was made by Jost Amman (1539-1591), and shows the vatman in the act of forming a sheet of paper upon a mould; also the stampers and press are visible in the background. A reproduction of the engraving, with a translation of the verse by the cobbler-poet, Hans Sachs, which appeared underneath, are given on this page. A facsimile of the title-page of this early book of trades is also shown.

Rags are brought unto my mill
Where much water turns the wheel,
They are cut and torn and shredded.
To the pulp is water added;
Then the sheets 'twixt felts must lie
While I wring them in my press.
Lastly, hang them up to dry
Snow-white in glossy loveliness.

Eygentliche Beschreibung
Aller Stände auff Er=
den / Hoher und Nidriger / Geistlicher
und Weltlicher / Aller Künsten / Handwercken
und Händeln / rc. vom grösten biß zum kleinesten /
Auch von jrem Ursprung / Erfindung und
gebreuchen.

Durch den weitberümpten Hans Sachsen

Gantz fleissig beschrieben / und in Teutsche Reimen ge-
fasset / Sehr nutzbarlich und lustig zu lesen / und auch mit künstreichen
Figuren / deren gleichen zuvor niemands gesehen / allen Ständen
so in diesem Buch begriffen / zu ehren und wolgefallen / Allen
Künstlern aber / als Malern / Goldschmiden / rc.
zu sonderlichem dienst in Druck
verfertigt.

Mit Röm. Keys. Maiest. Freyheit.
Gedruckt zu Franckfurt am Mayn.
M. D. LXVIII.

Title-page of Eygentliche Beschreibung aller Stände, etc.

GARZONI, TOMMASO, 1549?-1589.

La piazza vniversale di tvtte le professioni del mondo, e nobili et ignobill nvovamente formata, e posta in luce da Tomaso Garzoni da Bagnacauallo, mo al sereniss. et invittiss. Alfonso Secondo da Este Ducadiferrara. Venetia, 1586. 15 x 22 cm.

This volume, like other contemporary books of trades, contained short descriptions of occupations, and papermaking was also represented. An edition with German text was published at Frankfurt am Mayn in 1641.

LA

PIAZZA

VNIVERSALE

DI TVTTE LE PROFESSIONI

DEL MONDO, E NOBILI

ET IGNOBILI.

NVOVAMENTE FORMATA, E
posta in luce da Tomaso Garzoni
da Bagnacauallo.

AL SERENISS.MO ET INVITTISS.MO
ALFONSO SECONDO DA ESTE
DVCA DI FERRARA.

CON PRIVILEGIO.

IN VENETIA,
Appresso Gio. Battista Somascho.
M D LXXXVI.

The title-page of the 1586 edition of Garzoni's work.

CHURCHYARD, THOMAS, 1520?-1604.

A Description and playne Discourse of Paper, and the whole benefitts that Paper brings, with rehearsall, and setting forth in verse a Paper Myll built near Dartford, by an High Germaine, called Master Spilman, Jeweller to the Queenes Majestie. 1588.

This poem was the first account of papermaking in English and is interesting more for that reason than for any technical information that might be gained from it. The complete poem consisted of 353 lines printed in a quarto volume of seventeen leaves, with an engraved title-page. The work was prefaced by a sort of dedicatory epistle addressed "To my honorable friend Sir Walter Ralegh," under the title "A Sparke of Friendship and warm good will, etc."

There is an original copy of this most interesting work in the Bodleian Library, and a reprint may be found in Nichols, The Progresses of Queen Elizabeth, vol. ii., page 582. The poem is too long to give in its entirety here but a few salient verses may be quoted:

I prayse the man that first did Paper make,
 The only thing that sets all virtues forth:
It shooes new bookes, and keepes old workes awake,
 Much more of price than all the world is worth:
Though parchment duer a greater time and space,
 Yet can it not put paper out of place:
For paper, still, from man to man doth go,
 When parchment comes in few mens' hands you knowe.

Then he that made for us a Paper mill,
 Is worthy well of love and worldes good will.
And though his name be Spill-man by degree,
 Yet Help-man, now, he shall be calde by mee.
Six hundred men are set at worke by him,
 That else might starve, or seeke abroad their bread;
Who nowe live well, and goe full brave and trim,
 And who may boast they are with paper fed.

The hammers thump, and make as lowde a noyse,
 As fuller does that beates his woolen cloth,
In open shewe, then sundry secrete toyes,
 Makes rotten ragges to yeelde a thickned froth:
Then it is stampt, and washed as white as snowe,
 Then flong on frame, and hang'd to dry, I trow:
Thus paper streight it is, to write upon,
 As it were rubbe and smoothde with slicking-stone.

ZONCA, VITTORIO.

Novo teatro di machine et edificii per uarie et sicure operationi. Padoua, 1607. 19 x 27 cm.

The plate from the volume by Vittorio Zonca, depicting the stampers which were used for the maceration of linen and cotton cloth for papermaking.

PANCIROLLI, GUIDONIS.

Rerum Memorabilium sive Deperditarum, pars prima et secunda. Francofurtens, 1660. 16 x 20 cm.

The art of papermaking is treated in an insignificant way on pages 251-257 in the second part of this volume.

MEXIA, PEDRO, 1496-1552.

Silva De Varia Lección en la qual se tratan muchas cosas muy agradables y curiosas. Madrid, 1662. 17 x 22 cm.

An article on the origin of paper may be found on pages 306-310 of this work. Very few references to paper are found in early Spanish books.

BOECKLER, GEORGIUS ANDREAS.

Theatrum machinarum novum exhibens opera molaria et aquatica constructum industria. Nürnberg, 1662. 22 x 34 cm.

In this volume there are 154 full-page plates depicting various types of machinery, each engraving having a short explanation in Latin.

Plates 73 and 74 are devoted to paper and present the clearest delineation of the art to this date.

Plate 73, reproduced on page 19, shows the water wheel, stampers, vat, press, and finished paper hung to dry. The vatman is in the act of dipping the mould to form a sheet of paper, and another workman is operating the press. Bundles of rags may be seen in the foreground.

Plate 74 from the volume on machinery by G. A. Boeckler, published in Nürnberg in the year 1662.

Plate 74, reproduced on this page, gives a clear conception of the stampers which were used for the maceration of rags at the inception of papermaking in Europe. The Latin description of this plate by Boeckler reads: "On the river bank stands conveniently the water mill wheel which turns an axle fitted with cams. These various cams in turn operate the stamps, lifting them up and letting them fall full force upon the rags. Through the trough water from the river is distributed to the receptacle which serves to hold the rags."

It is evident that the engraver who made the plate for Boeckler used the earlier picture from the volume of Vittorio Zonca (page 17) for the model.

Plate 73 from the book by G. A. Boeckler, showing the process of papermaking.

FULLER, THOMAS, 1608-1661.

The history of the worthies of England. Endeavoured by Thomas Fuller, D. D. London, 1662. 24 x 35 cm.

This volume is arranged by counties, with the commodities, manufactures, etc., of each, followed by brief biographies of its "worthies."

Fuller gives the following observations regarding the paper of his time: "Paper participates in some sort of the characters of the country which make it; the Venetian being neat, subtile, and court-like; the French, light, slight, and slender; and the Dutch, thick, corpulent, and gross, sucking up the ink with the sponginess thereof. To such who object that we can never equal the perfection of Venice paper, I return, neither can we match the purity of Venice glasses; and yet many green ones are blown in Sussex, profitable to the makers, and convenient to the users. Our home-spun paper might be found beneficial."

FRAME, RICHARD.

A short description of Pennsilvania, Or, A Relation of What things are known, enjoyed, and like to be discovered in the said Province. Presented as a token of Good Will to the people of England. Philadelphia, 1692.

In this book may be found the following poem regarding the Rittenhouse paper mill, established near Germantown, Pennsylvania, about 1690:

The German-Town, of which I spoke before,
Which is, at least in length one Mile or more,
Where lives High-German People, and Low-Dutch,
Whose Trade in weaving Linen Cloth is much.
There grows the Flax, as also you may know,
That from the same they do divide the Tow;
Their trade fits well within this Habitation,
We find Convenience for their Occasion.
One Trade brings in employment for another,
So that we may suppose each Trade a Brother;
From Linen Rags good Paper doth derive,
The first Trade keeps the second Trade alive;
Without the first, the second cannot be,
Therefore since these two can so well agree,
Convenience doth appear to place them nigh,
One in German-Town, t'other hard by.
A Paper-Mill near German-Town doth stand.
So that the Flax, which first springs from the Land,
First Flax, then Yarn, and then they must begin
To weave the same, which they took pains to spin.
Also, when on our backs it is well worn,
Some of the same remains Ragged and Torn;
Then of the Rags our Paper it is made,
Which in process of time dost waste and fade:
So, what comes from the Earth, appeareth plain,
The same in Time, returns to Earth again.

MANDEY, VENTERI.

Mechanick Powers; or, The Mystery of Nature and Art unvail'd. Shewing what Great Things may be perform'd by Mechanick Engines, in removing and raising Bodies of vast Weights: with little Strength or

Force; and also the making of Machines or Engines, for raising of Water, draining of Grounds, and several other Uses. London, 1696. 15 x 20 cm.

This volume gives the first description in English of the appliances used in the maceration of papermaking material. One reference (page 71) reads: "....Also Engines to make Paper, in which a wooden Cylinder of sufficient magnitude standing out, furnisht with little troughs, takes up wooden Mallots, which bruise and break whatever is laid under them to powder."

Mechanick Powers:

OR, THE

MYSTERY

OF

NATURE *and* ART

UNVAIL'D

Shewing what *Great Things* may be perform'd by Mechanick Engines, in removing and raising Bodies of vast Weights: with little Strength or Force; and also the making of Machines or Engines, for raising of Water, draining of Grounds, and several other Uses.

TOGETHER,

With a Treatise of *Circular Motion* artificially fitted to Mechanick use, and the making of Clock-work, and other Engines.

A Work pleasant and profitable for all sorts of Men, from the highest to the lowest Degree: And never treated of in *English* but once before, and that but briefly.

The whole comprized in Ten Books, *and illustrated with Copper Cuts.*

By *Ven. Mandey*, Philomat.

LONDON,

Printed for *Tho. Shelmerdine* at the *Rose tree*, and *Tho. Ballard*, at the *Rising-Sun* in *Little-Brittain*, 1702. Where is to be Sold, the *Universal Mathematical Synopsis*, by the same Author. Price 6 *s.*

The title-page of the second edition of the book on mechanics by Venteri Mandey. This edition was issued in 1702 but the text and engravings were identical with the 1696 edition.

HOLME, JOHN.
A True Relation of the Flourishing State of Pennsilvania. 1696.

The following poem by John Holme is supposed to have been the first metrical composition compiled in Pennsylvania, but it was not published until 1847. The poem related to William Bradford and the Rittenhouse mill:

Here dwelt a printer and I find
That he can both print books and bind;
He wants not paper, ink, nor skill
He's owner of a paper mill.
The paper mill is here hard by
And makes good paper frequently,
But the printer, as I do here tell,
Is gone into New York to dwell.
No doubt but he will lay up bags
If he can get good store of rags.
Kind friend, when thy old shift is rent
Let it to th' paper mill be sent.

THOMAS, GABRIEL.
An Historical and Geographical Account of the Province and Country of Pennsilvania. London, 1698.

In this book Thomas says: "All sorts of very good paper are made in German-Town, as also very fine German linen such as no person of quality need be asham'd to wear."

PAPPIERMACHER-KUNST.
Die Preiss-würdige Pappiermacher-Kunst, Oder: Gründliche Beschreibung, Was für eine edle Kunst und theure Gabe GOTTES es sey umb die höchst-nöthige Pappiermacher-Kunst, wie es aus GOttes Wort und denen Historien, wie auch andern dienlichen Gründen, mit der täglichen Praxi und Erfahrung, ec. zu beweisen. Hamburg, 1699. 11 x 17 cm.

This little book of 96 pages is one of the rarest of papermaking items. A facsimile of the title-page of this volume is given on page 23.

WANLEY, HUMPHREY, 1671-1726.
A letter of papermaking interest was written in 1699 by Wanley to Samuel Pepys, the diarist. Wanley was custodian of heraldic manuscripts

to Robert Harley, Earl of Oxford. The letter is in regard to a collection of paper Wanley had formed and reads as follows: ". . . . This collection of shreads and scraps of old parchment makes me bold to mention to you, Sir, another like collection of paper that I have made when I was last in London. This I have put together in two volumes of which I venture to give you some account. (These two volumes, bound in one book, now form the Sloane Manuscript No. 526, in the British Museum.)

"The first volume contains: Some of our countrey common brown, grey, blue, red, and black paper. Some Eastern paper written upon in the East Indies, Persia, Turkey, and China, and some Chinese paper printed upon. . . .

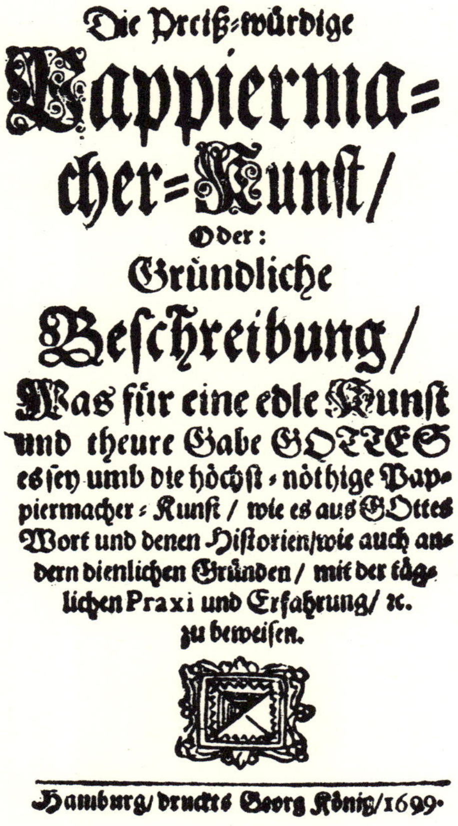
Die Preiß-würdige
Pappiermacher-Kunst/
Oder:
Gründliche
Beschreibung/
Was für eine edle Kunst
und theure Gabe GOTTES
es sey umb die höchst-nöthige Pappiermacher-Kunst/ wie es aus GOttes
Wort und denen Historien/wie auch andern dienlichen Gründen/ mit der täglichen Praxi und Erfahrung/ rc.
zu beweisen.
Hamburg/ druckts Georg König/1699.

The title-page of Pappiermacher-Kunst, a rare book on papermaking, described on page 22.

"The other volume contains a collection of sheets and pieces of white European paper which are plac'd according to their several marks and stamps. The reason why I gathered them together was not only to satisfy my curiosity, in knowing what sort of paper is in use among so many nations, or to please myself or my friends with looking on the fine colours, etc., but to know whether they would confirm a notion I have of the originall of our paper, which, to me, they do. And further, I thought that some time or other the history of paper might be attempted, and with so much the more hope of success, when the author, whoever he may be, has before his eyes most of the sorts of that subject that he is to treat about."

BAGFORD, JOHN, 1675-1715.

In 1707 Bagford published in the Philosophical Translations: "A proposal for a history of printing, printers, illuminators, and papermaking; on subscription 10 shillings, and 10 shillings more on delivery of a volume in folio, containing two hundred sheets." This book was never published but the material which Bagford amassed is now in the British Museum, bearing the numbers 5891 to 5988. In volume 5940 there are specimens of end-papers and a number of watermarks from the sixteenth century. In volume 5941 there may be seen samples of Oriental papers and in volume 5942 there are examples of paper with imprints of the makers and printed advertisements relative to the manufacture of paper.

John Bagford was considered the best authority on paper of his time.

VAN SAINT CLARA, ABRAHAM, 1644-1709.
Iets Voor Allen, Zynde een Verhandeling en Verbeelding Van allerhande. Standen, Studien, Konsten, Wetenschappen, Handwerken, etc. Amsterdam, 1717. 10 x 16 cm.

IETS
VOOR ALLEN,
Zynde een
VERHANDELING
en
VERBEELDING
Van allerhande.
STANDEN, STUDIEN, KONSTEN,
WETENSCHAPPEN,
HANDWERKEN enz:
Waar in derselver Oorsprong, Opkomst, Aardt,
Eygenschap, Deugdt en Ondeugden, op een
Geestige wyze beschreven werdt.
In het Hoogduyts zamengesteldt, en alom met
Vreemde geschiedenissen, Zinryke Zinbeelden
en overaardige gedagten doormengt,
Door den beruchten PATER
ABRAHAM VAN St. CLARA.
In 't Nederduyts overgebragt
DOOR
J. LE LONG.

t' AMSTELDAM.
By de JANSSOONS VAN WAESBERGE.
1717.

Title-page of the St. Clara work

This small Dutch book of trades devotes five pages to the art of papermaking and presents a well-executed engraving (reproduced on this page) of the vatman forming a sheet of paper.

The engraving from the St. Clara book of trades.

Saint Clara was a Roman Catholic preacher of Vienna and his earliest writing regarding paper appeared about the year 1700 under the caption, "Was ist ein Papier?" The verse reads:

"Ein Papier ist ein Werkzeug der Gelehrten, ist ein Vorrat der Bücher. Ein Papier ist ein Hausrat der Kanzlei, ist ein Schatz der Schüler, eine Erhalterin der menschlichen Freundschaft; ein Papier ist der grösste Nutz der Welt. Mein Papier, wer sind, bekenne es recht, deine Eltern gewest? Mein Vater, sagtes, war ein Lumpen, meine Mutter ein Fetzen. Von einem so schlechten Lumpengesind eine so herrliche Sach!"

LUIKEN, JAN EN KASPER, 1670-1710.

Spiegel van het Menselyk Bedryf, vertoonende honderd verscheiden Ambachten, Konstig afgebeeld, en met Godlyke Spreuken en Stitchtlyke Verzen verrykt. Amsterdam, 1718. 10 x 16 cm.

In this Dutch book of trades there are one hundred engravings, one of which represents the papermaker. This volume was founded on the work of Abraham Van Saint Clara which was published in 1717. The only difference in the two plates depicting the papermaker is that in the later edition the picture was reversed, showing it had been traced from the engraving of 1717.

STURM, LEONHARDT CHRISTOPH, 1669-1719.

Vollständige Mühlen Baukunst. Augsburg, 1718. 23 x 35 cm.

The earliest engraving of the Hollander, or beating-engine, may be found in this rare volume on machinery and mill appliances. Beating-engines were introduced about the year 1700 and were used for the maceration of rags for papermaking. These machines superseded the ancient stamping-mills as pictured by Zonca and Boeckler and reproduced on pages 17 and 18 of this bibliography.

The Hollanders were usually placed in a circle in the mills so that four or five of the machines were operated by one huge wooden wheel, fitted with stout round pegs which acted as cogs.

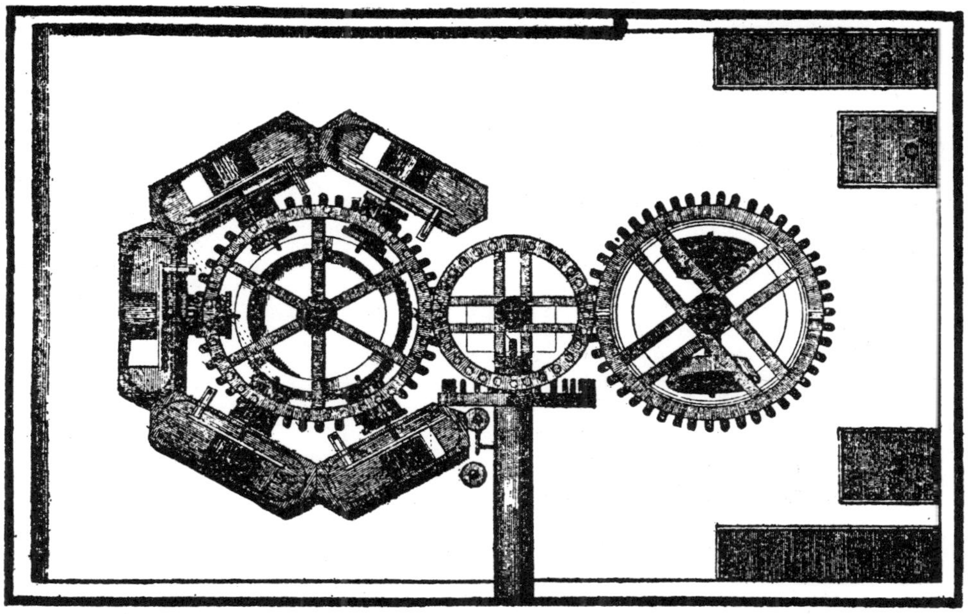

The earliest engraving of the Hollander.
From Vollständige Mühlen Baukunst.

In the engraving are pictured five Hollanders and two mill-stones. The mill-stones were used for grinding grain; at the same time the beaters were reducing rags to pulp for making into paper. This rather incongruous combination was seldom used by the Dutch papermakers and the placing of the mill-stones in the print was little more than a phantasy of Sturm, the designer, who executed the drawing after a tour of inspection of the paper mills of Holland.

This volume by Leonhardt Christoph Sturm is one of the rarest and most important books of the eighteenth century that treats of papermaking.

RÉAUMUR, RENÉ ANTOINE FERCHAULT DE, 1683-1757.

A pamphlet compiled about 1720 is credited to Réaumur in which he advanced the theory that wood might be used as a papermaking material. He had observed that the fabric of wasps' nests was procured from wood and suggested that some one with opportunity make the experiment to ascertain if paper could be made from the same material. It has not been possible to locate a copy of this pamphlet in any of the great American or European Libraries.

NOVUM ITALIAE THEATRUM

sive accurata Descriptio ipsius urbium, palatiorum, sacrarum aedium, etc. Hagae-comitum, 1724. 4 volumes. 35 x 47 cm.

These volumes give descriptions of the cities, palaces, and sacred buildings of Italy and the only relation to the subject of papermaking is in volume 2., plate xxvii, which shows an interior view of a Fabriano paper mill.

The engraving in Novum Italiae Theatrum, depicting an interior view of an old Fabriano paper mill.

BRUECKMANN, FRANZ ERNST, 1697-1753.

Magnalia Dei in locis svbterraneis, oder Unterirdische schatz-cammer aller königreiche und länder, in ausführlicher beschreibung aller, mehr als MDC. bergwercke durch alle vier welt-theile . . . nebst anmerckung aller derjenigen länder und oerter, wo edelgesteine zu finden, in geographischer ordnung und einigen kupffer-figuren zu besichtigen dargestellet von Francisco Ernesto Brvckmann.
Braunschweig, 1727-30. 2 volumes. 22 x 35 cm.

Four copies of this work on geology were supposed to have been printed upon paper made from asbestos, the first paper ever fabricated from this material.

SEBA, ALBERT, 1665-1736.

Locupletissimi rerum naturalium thesauri accurata descriptio, et iconibus artificiosissimis expressio, per universam physices historiam. Opus, cui, in hoc rerum genere, nullum par exstitit. Ex toto terrarum orbe collegit, digessit, descripsit, et depingendum curavit Albertus Seba.
Amstelaedami, 1734-65. 4 volumes. 35 x 53 cm.

Seba, a Flemish writer on natural history, called attention to the fact that his country "does not seem to lack trees suited for making paper, if people would give themselves the necessary expense and trouble. 'Alga marina' for example, which is composed of long, strong, viscous filaments, might it not be proper for this purpose, as well as the mats of Muscovy, if they were prepared as the Japanese make their timber?"

VAN NATRUS, LEENDERT.

Groot Volkomen Moolenboek; of, Naauwkeurig Ontwerp Van Allerhande tot nog toe bekende Soorten van Moolens, Met haare Gronden en Opstallen, en al het geene verder daar toe behoort. Ten dienste der Oeffenaaren van deze Kunst, en van alle Liefhebbers der Bouw- en Tekenkunst, geteekent door Leendert Van Natrus, Moolemakers Baas van de Ed: Oost-Indische Compagnie, te Amsterdam, Jacob Polly, Moolemaker te Saardam; en Cornelis Van Vuuren; En zeer naauwkeurig in het Kooper gebragt door Jan Punt.
Amsterdam, 1734-35. 2 volumes in one. 33 x 54 cm.

Groot Volkomen Moolenboek is a rare treatise on machinery and the equipment of windmills. Plates number 16, 17, 18, and 19 relate to paper manufacturing and descriptive text is given on pages six and seven. Plate 16 outlines the general floor plan of a paper mill with the arrangement of the various appliances that were used in making paper. In plate 17 the construction of the presses is shown, also the distribution of the hair ropes upon which paper was hung to dry. Plate 18 (reproduced below) gives en-

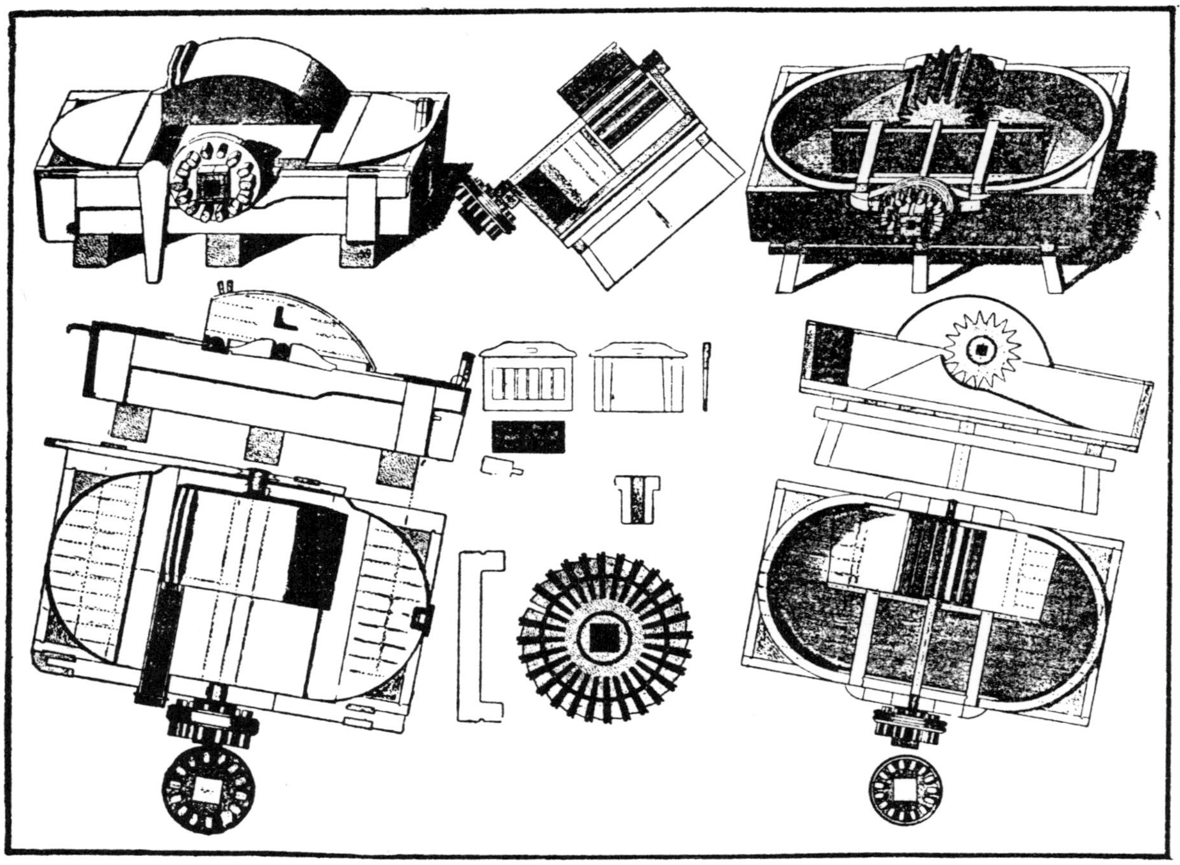

The first Dutch engraving of the Hollander. From Groot Volkomen Moolenboek, Amsterdam, 1734-36.

gravings of Hollanders, or beating engines, which were used for the preparation of papermaking material. This was the first presentation of these machines in the country of their invention. Plate 19 depicts a windmill with the Hollanders in such a position that the maximum amount of rags could have been beaten by the limited power of the wind.

LEWIS, JOHN, 1675-1747.

The Life of Mayster Wyllyam Caxton, of the Weald of Kent; the First Printer in England. In which is given An Account of the Rise and Progress of the Art of Pryntyng in England, during his Time, till 1493. London, 1737. 13 x 22 cm.

This was the earliest English book to take notice of watermarks. Two full-page plates are devoted to reproductions of twenty papermarks that were used in the paper upon which the books of Caxton were printed, from 1477 to 1490. The marks include such devices as shields, animals, shears, bulls' heads; also a crown, a hand, a ship, and the letter P, which is common in Caxton's books, probably the initial of Philip the Good.

SCHWARZ, CHRISTIAN GOTTLIEB.

Primaria qvaedam docvmenta de Origine Typographiae qvorvm illvstratorvm partem primam svb praesidio Christiani Gottlibii Schwarzii. Altorfii, 1740. 15 x 17 cm.

A number of early watermarks are reproduced in this volume which constitute the only papermaking interest in the book.

PRIMARIA QVAEDAM DOCVMENTA
DE
ORIGINE TYPOGRAPHIAE
QVORVM ILLVSTRATORVM
PARTEM PRIMAM
SVB
PRAESIDIO
CHRISTIANI GOTTLIBII
SCHVVARZII
COM. PAL. CAES. ET PROF. PVBL.
A. D. XII NOVEMBRIS A. S. R. cIↃIↃ CCXXXX
DISQVISITIONI ACADEMICAE
SVBIICIT
BENEDICTVS GVILIELMVS MVNCH
NORIMBERGENSIS

IPSO ANNO TERTIO SAECVLARI TYPOGRAPHIAE
DIVINO AVXILIO A GERMANIS INVENTAE.

ALTORFII
TYPIS IOHANNIS ADAMI HESSELII.

Title-page of the volume on typography by Schwarz.

GUETTARD, JEAN ÉTIENNE, 1715-1786.

Observations sur differentes matières dont on fabrique le papier. In: Memoires de Paris, 1741.

Apparently the first notes written by Guettard on the subject of paper.

PANSA, MUTIO.

An Account of the Original of Writing and Paper. In: The Harleian Miscellany. Volume 3, pages 336-9, 1744.

This article was taken from a book entitled La Libraria Vaticana written by Mutio Pansa, keeper of the Vatican Library, and printed in Rome. The Miscellany gives from this volume: "Of the Use of Books and the Invention of Letters, and of the Paper of the Ancients, of the Papyrus of the Romans . . . and the Paper of our Times."

PAPERMAKING TRADE.

A general description of all Trades, digested in alphabetical order: by which Parents, Guardians, and Trustees, may, with greater ease and certainty, make choice of Trades agreeable to the Capacity, Education, Inclination, Strength, and Fortune of the Youth under their Care. London, 1747. 10 x 17 cm.

This was the first English book of trades to include papermaking as a suitable calling or avocation for the youth of England. The section of the book devoted to this craft reads: "Papermaking requires much water and a great deal of room, and therefore is altogether carried on at water-mills in the country, which undertakings are not numerous. It goes through various operations and divers hands before it is completed, and the moulding part, which is the principal, requires a nice hand, and a good eye.

"They take with an apprentice 5 or 10 pounds, who work from six to six, and he ought not to be a very tender lad; they pay a journeyman moulder 15 or 18 shillings a week; and a mill with the proper utensils (beside which there must be a reserve of at least 200 pounds cash for a stock of rags, etc.) will stand one, minded to be a master, in 100 pounds of which some have two or three."

AMES, JOSEPH, 1688-1758.

Typographical Antiquities: Being an Historical Account of Printing in England: with some Memoirs of our Ancient Printers, and a Register of Books printed by them, from the Year mcccclxxi to the Year mdc. London, 1749. 19 x 26 cm.

The only papermaking interest in this volume is the reproduction of a number of watermarks from the paper upon which Caxton's books were printed. These facsimiles appear on pages 74 and 75 of Ames' work and are identical with those reproduced in the book by John Lewis, described on page 28 of this bibliography.

GUETTARD, JEAN ÉTIENNE, 1715-1786.

Mémoire sur les différentes manières de fabriquer le papier. In: Histoire de l'Académie des sciences, année 1741; Journal économique. July and August, 1751.

In this account of the different ways of manufacturing paper Guettard mentions the possibility of Conferva being used as a papermaking material. The account contains actual specimens of paper made from the bark, leaves, and wood of various trees, plants, and shrubs.

GUETTARD, JEAN ÉTIENNE, 1715-1786.

Inquiry concerning the materials that may be used in making paper. By Mr. Guettard, of the Royal Academy of Sciences, and Physician to his Serene Highness the Duke of Orleans. In: Select Essays on Commerce, Agriculture, Mines, Fisheries, and other useful subjects.
London, 1754. 14 x 23 cm.

On pages 372-406 of this volume on miscellaneous subjects Guettard recommends the use of various vegetation for papermaking. The French physician had not at this time carried his ideas to any practical conclusion for on page 406 of his essay he says: "Such examinations have always constituted my desires, since I thought of making experiments upon paper. I have not been able hitherto, to accomplish my scheme; but nevertheless, I have reason to hope I shall one day see it accomplished."

BASKERVILLE, JOHN, 1706-1775.

Publii Virgilii Maronis Bucolica, Georgica, et Æneis.
Birminghamiae, 1757. 23 x 29 cm.

Baskerville, the Birmingham printer, originated wove paper about the year 1750 and his edition of Virgil, issued in 1757, was the first book in which paper of this kind was used. However, only about half of the paper in each volume of the edition is wove, the remainder being of the laid type.

FABRICIUS, JOHANN ALBERT, 1668-1736.

Bibliographia Antiquaria, sive Introductio in Notitiam Scriptorum, qui Antiquitates Hebraicas, Graecas, Romanas et Christianas Scriptis illustraverunt. Editio Tertia. Hamburgi, 1760.

Several of the early authors on the subject of ancient writing substances are recorded in this volume.

VAN ZYL, JOHANNIS.

Theatrum Machinarum Universale; of Groot Algemeen Moolen-Boek, Behelzende de Beschryving en Afbeeldingen van allerhande soorten van Moolens . . . . En in 't Koper gebragt door Jan Schenk.
Nieuwe Beschryving van Gronden en Opstallen van verscheide Moolens.
Amsterdam, 1761.
38 x 54 cm.

This book, founded on the work of Leendert Van Natrus, is an important treatise on the construction of windmills and equipment.

There are six engravings devoted to the art of papermaking, with four pages of explanatory text giving an entire account of the craft as it was practiced at that time.

It was from the study of this comprehensive volume and the works of such writers as Sturm and L. Van Natrus, the European papermakers of the eighteenth century derived their technical knowledge.

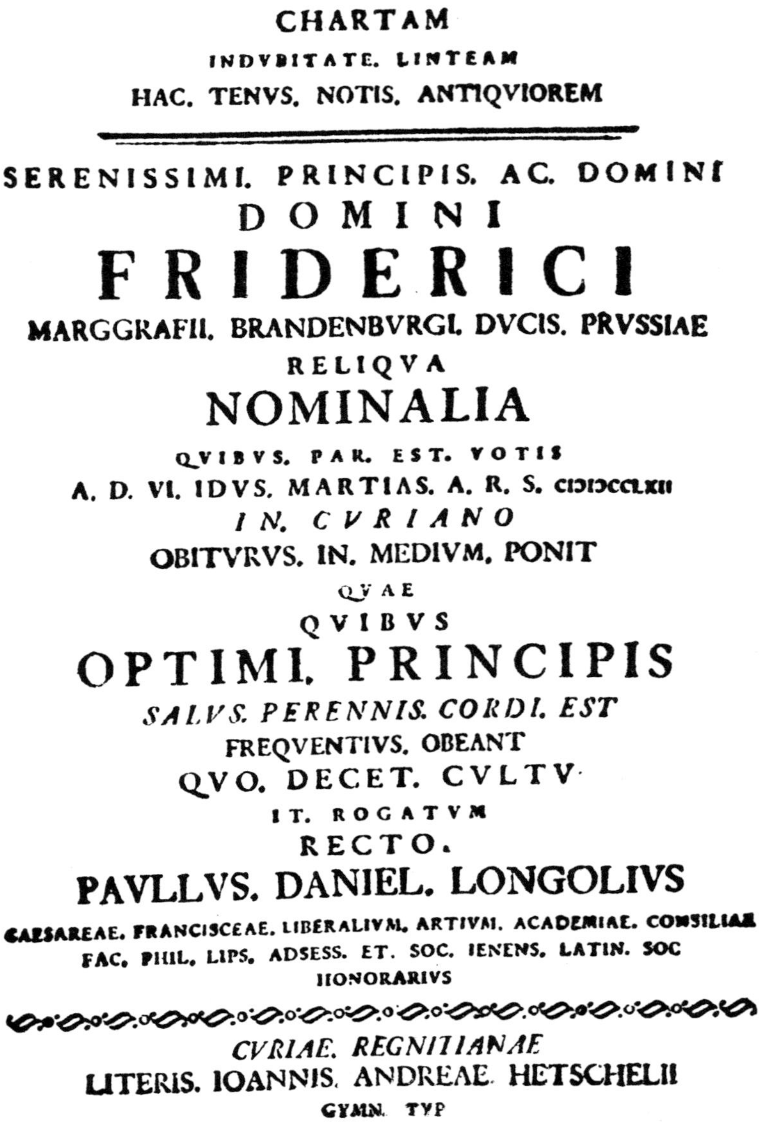
CHARTAM
INDVBITATE. LINTEAM
HAC. TENVS. NOTIS. ANTIQVIOREM

SERENISSIMI. PRINCIPIS. AC. DOMINI
DOMINI
FRIDERICI
MARGGRAFII. BRANDENBVRGI. DVCIS. PRVSSIAE
RELIQVA
NOMINALIA
QVIBVS. PAR. EST. VOTIS
A. D. VI. IDVS. MARTIAS. A. R. S. CIƆIƆCCLXII
IN. CVRIANO
OBITVRVS. IN. MEDIVM. PONIT
QVAE
QVIBVS
OPTIMI. PRINCIPIS
SALVS. PERENNIS. CORDI. EST
FREQVENTIVS. OBEANT
QVO. DECET. CVLTV
IT. ROGATVM
RECTO.
PAVLLVS. DANIEL. LONGOLIVS
CAESAREAE. FRANCISCEAE. LIBERALIVM. ARTIVM. ACADEMIAE. CONSILIAR
FAC. PHIL. LIPS. ADSESS. ET. SOC. IENENS. LATIN. SOC
HONORARIVS

CVRIAE. REGNITIANAE
LITERIS. IOANNIS. ANDREAE. HETSCHELII
GYMN. TYP

Title-page of the volume by Paul Daniel Longolius, concerning linen paper.

LONGOLIUS, PAUL DANIEL, 1704-1779.

Chartam indvbitate, linteam hac. tenvs. notis. antiqviorem serenissimi. principis, ac, domini Domini Friderici Marggrafii, Brandenbvrgi, Dvcis. Prvssiae Reliqva Nominalia qvibvs. par. est. votis A. D. VI, Idvs, Martias, A. R. S. MDCCLXII . . . Rector Pavllvs. Daniel. Longolivs.
Cvriae, Regnitianae, 1762.

LALANDE, JOSEPH JÉRÔME LE FRANÇAIS DE, 1732-1807.
Art de faire le papier. In: Description des arts et métiers, published by the Académie des sciences. Volume 4. Paris, 1761. 29 x 45 cm.

This is one of the most important technical works of the eighteenth century on papermaking and is more complete than any book of the period. There are fourteen large engravings (one folded) showing the entire process of the craft from sorting and cutting the rags to wrapping the finished paper. One hundred and fifty pages of text are devoted to describing the plates.

MEERMAN, GERARD, 1722-1771.
Syndici Roterodamensis, Admonitio de Chartae nostratis, seu lineae, origine. Rotterdam, 1762.

Meerman, a Hollander, offered a premium of twenty-five ducats for the discovery of the time linen paper was first fabricated. Many specimens were sent to him from all parts of Europe which were supposed to be early linen paper, but researches did not establish any definite conclusion regarding the origin of linen as a material for paper.

LEWIS, WILLIAM, 1714-1781.
Commercium philosophico-technicum; or, The philosophical commerce of arts: designed as an attempt to improve arts, trades, and manufactures. London, 1763. 17 x 26 cm.

On page 393 of this volume will be found a short treatise on the subject of paper under the heading: "Of the preparation of the paper for durable writing."

STRANGE, JOHN, 1732-1799.
Letter on the origin of the natural paper of Cortona, with other observations relative to the uses and excellent qualities of the Conferva of Pliny. Pisa, 1764.

John Strange, an English diplomatist and traveler, wrote this description of papermaking while visiting Italy. He tells about the paper of Cortona, a town of Tuscany, and gives a summary of the use of broom, a fabaceous shrub, and other plants for papermaking. While John Strange recommends Conferva for various uses he does not deem it suitable material from which to make paper. The researches made by Strange are vague and unimportant.

SCHAEFFER, JACOB CHRISTIAN, 1718-1790.

Versuche und Muster ohne alle Lumpen oder doch mit einem geringen Zusatze derselben Papier zu machen.

Regensburg, 1765-71. 6 volumes. 82 paper specimens. 13x18 cm.

This is the most interesting and rarest work on the subject of paper ever published, but five complete copies being known to exist in America.

Jacob Christian Schäffers

Doctors der Gottesgelehrsamkeit und Weltweisheit; Ev. Pred. zu Regensburg; Sr. Königl. Maj. zu Dännemark Norwegen Rathes und Prof. honor. zu Altona; der Academie der Naturforscher, zu Petersburg, Londen, Berlin, Upsal, Roveredo, München, und Mannheim, der Gesellschaft der Wissenschaften zu Duisburg, physischbotanischen Gesellschaft zu Florenz und historischen Gesellschaft zu Göttingen; der deutschen Gesellschaft zu Leipzig, Göttingen, Altdorf und Erlangen Mitgliedes; wie auch der Academie zu Paris Correspondenteus

Neue

Versuche und Muster

das

Pflanzenreich

zum

Papiermachen und andern Sachen

wirthschaftsnützlich zu gebrauchen.

Erster Band.

Nebst zwo ausgemahlten Kupfertafeln.

Regensburg, 1765.

Title-page of the second part of Schaeffer's work.

Jacob Christian Schaeffer was born in Querfurt, near Merseburg in Saxony in the year 1718, but his family early removed to Regensburg, or Ratisbon, the capital of the Bavarian province of Oberpfalz, about one hundred and fifty miles south of his birthplace.

Schaeffer studied for the ministry and became a well-known clergyman in Regensburg, but it seems he derived his greatest pleasure outside the church as he devoted most of his time to the study of science and natural history. His work on the Fungi of Bavaria is still regarded as a standard authority. Schaeffer's interest in the flora of Bavaria directed his attention to the possibility of new materials for papermaking and it is with his researches in this subject that we are concerned.

The frontispiece to the first volume of the rare work by Jacob Christian Schaeffer.

With his six volume treatise Dr. Schaeffer left a permanent record of his experiments for new papermaking materials, and the actual specimens of paper in his volumes establish the fact that he was the pioneer in the use of many vegetable fibres for the fabrication of paper.

It was not Schaeffer's desire to make well finished paper; he simply wished to show the vast variety of vegetation available for the purpose. As his experiments were carried on previous to the discovery of bleach the samples of paper are the tint of the original material from which they were made.

In most of the examples about one-fifth part cotton rags were added to the pulp to help bind the fibres together. A number of the specimens are sized and nearly all have been printed upon.

It is curious to note one of the first specimens shown in Schaeffer's books was made from wasps' nests—for was not the wasp, himself, the first papermaker, or was it the frog who was the original fabricator of paper? The wasp made his nest of wood fibre cleverly felted together exactly as paper is constructed, while the frog made a peculiar kind of spittle on the surface of ponds which became well-felted paper after drying naturally in the sun.

Dr. Schaeffer's researches in papermaking extended over a period of more than eight years and all the materials with which he experimented were gathered in his own garden or neighboring fields. The mode of making the various samples of paper is given in detail. The materials, according to their nature, were first chopped by hand, the different kinds of wood having been reduced by a toothed plane. Dr. Schaeffer carried on most of the experiments in his own home, and for the purpose of beating the materials to pulp he had a miniature set of stampers which was operated by hand. In most cases the vegetable matter was subjected to the action of the stampers without any preliminary preparation, while with a few of the materials milk of lime was used. With some of the more stubborn fibres Dr. Schaeffer employed a stiff lime paste and suffered the material to remain in it for various lengths of time. He noted that this treatment reduced the beating process and also that the lime gave a brown tint to the wood.

Watermark of the crest.

All the samples of paper were formed on a laid mould which measured about eleven by fifteen inches, and an indistinct watermark of a crest appears on most of the specimens. The paper upon which the text of the volumes is printed bears a watermark of crossed-keys with the initials C M M inside a border.

There are not many copies of Schaeffer's work extant and the specimens in the various sets are not entirely identical. The compilation is more often found bound complete in one volume, than in the six original pamphlets as probably issued by Dr. Schaeffer.

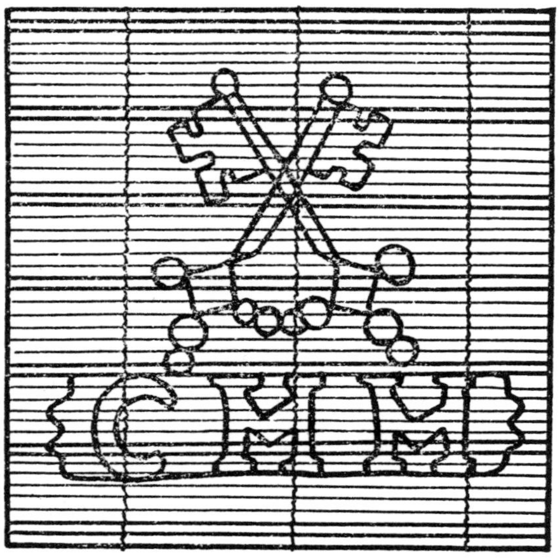

The watermark of the crossed-keys.

The first volume of this valuable work was completed January 30th 1765 and the book contains fifty-five pages of text, five plates, and fifteen specimens of paper. The examples include papers made from wasps' nests and various kinds of wood, moss, and vines.

In volume two, dated April 3d 1765, there are twenty-eight pages of text and one plate. The specimens include papers fabricated from hemp, bark, straw, and cabbage stalks.

The third volume, dated November 3d 1765, has thirty-two text pages and the specimens embrace papers made from asbestos, cattail and burdock stalks, thistles and turf.

Volume four was finished January 1st 1766 with twenty-four pages of text, and the eleven examples include papers from seed, mallow, St. John's wort, and Indian corn husks.

The fifth volume is dated April 15th 1767, while the date of the sixth volume is 1771. These two books give specimens of paper made from genista, pine-cones, potatoes, old shingles, reeds, and bean, horse-chestnut, walnut, tulip, and linden leaves; also paper made from yellow and brazil wood.

Vollständige
Mühlen Baukunst

Darinnen werden

I. Alle Grundregeln so zu der PRAXI nöthig, die doch gar wenigen recht bekant sind, treulich angewiesen;

II. Die Vortheile, die man bey Anlegung der Wasserräder alle Sorten von Machinen zutreiben in acht nehmen muß, Auf den höchsten Grad der Vollkommenheit gebracht;

III. Was insonderheit an Korn- Graupen- Papier- Öhl- Pulfer- Säg- Steinschneide- Bohr- Schleiff- Sensen- Kessel- Eisendrat- Hächsel- und Dreschmühlen zuverbessern, aufrichtig entdecket.

Also, daß dieses Werck wohl vor eine Entdeckung der aller raresten und vortrefflichsten Mechanischen Vortheile in der Praxi darff angegeben werden.

Cammer- und Policey-Räthen, Beamten, Statt-Magistraten, Kauffleuthen und allen sowohl der grossen Oeconomie als Mechanischer Künste beflissenen zum Nutzen getreulich eröffnet

von

Leonhardt Christoph Sturm.

AUGSPURG
Verlegt von Jeremias Wolff Kunsthändlern, A°. 1718.
Cum Privilegio Sac. Cæs. Majestatis.

GROOT VOLKOMEN
MOOLENBOEK;

OF

NAAUWKEURIG ONTWERP

Van Allerhande tot nog toe bekende Soorten van

MOOLENS,

Met haare GRONDEN en OPSTALLEN,

en al het geene verder daar toe behoort.

Ten dienste der Oeffenaaren van deze Kunst, en van alle Liefhebbers der Bouw- en Tekenkunst, geteekent

DOOR

LEENDERT VAN NATRUS,

Moolenmaakers Baas van de Ed: Oost-Indische Compagnie, te Amsterdam,

JACOB POLLY,

Moolemaker te Saardam;

EN

CORNELIS VAN VUUREN,

En zeer naauwkeurig in het Kooper gebragt

Door JAN PUNT.

EERSTE DEEL.

Te AMSTERDAM,

By JOHANNES CÓVENS EN CORNELIS MORTIER.

MDCCXXXIV.

THEATRUM
MACHINARUM UNIVERSALE;

OF

GROOT ALGEMEEN
MOOLEN-BOEK,

Behelzende de BESCHRYVING en AFBEELDINGEN van allerhande ſoorten van MOOLENS, der zelver OPSTALLEN, en GRONDEN:

Waar onder veele, die eerſt onlangs zyn gebouwt, ſtaande zo buiten de Stad Amſterdam, als in de omleggende Plaatſen; nevens eene naauwkeurige Verbeelding in verſcheide Platen van de AMSTERDAMSCHE MODDER-MOOLENS; ook een nette Verdeeling van de Staven en Kammen, en meer andere zeer fraaie Werken ten nutte der BOUW- en TEKEN-KONST; nooit zo volkomen in 't licht gebragt.

Voor 't grootſte gedeelte getekent te Amſterdam

DOOR

JOHANNIS VAN ZYL,

Moolenmaker van Lexmondt;

En in 't Koper gebragt door

JAN SCHENK.

I DEEL.

Te AMSTERDAM,
By PETRUS SCHENK, Konſt- en Kaartverkoper, in de Kalverſtraat, by de Gaperſteeg, in N. Viſſchers Atlas. MDCCLXI.

The hand-operated stampers employed by Dr. Schaeffer in his experiments in reducing various materials to pulp.

MEERMAN, GERARD, 1722-1771.

Meerman et doctorum virorum ad eum epistolae atque observationes de chartae vulgaris seu lineae origine. Hagae-Comitum, 1767. 11 x 18 cm.

This was the second compilation by Gerard Meerman on the origin of linen paper.

GERARDI MEERMAN

ET DOCTORUM VIRORUM
AD EUM EPISTOLAE
ATQUE OBSERVATIONES

DE

CHARTAE VULGARIS SEU
LINEAE ORIGINE.

EDIDIT

AC PRAEFATIONE INSTRUXIT

JACOBUS VAN VAASSEN.

HAGAE-COMITUM,
Apud NICOLAUM VAN DAALEN.
MDCCLXVII.

The title-page of Gerard Meerman's book on the origin of linen paper.

GUETTARD, JEAN ÉTIENNE, 1715-1786.

Recherches sur les matières qui peuvent servir a faire du papier. In: Mémoires sur différentes parties des Sciences et Arts. Volume 1. Paris, 1768.

SCHAEFFER, JACOB CHRISTIAN, 1718-1790.

Proefnemingen en monsterbladen, om papier te maaken zonder lompen, of met een gering Byvoegzel derzelven. Amsterdam, 1770. 14 x 19 cm.

The rare Dutch edition of the work on papermaking materials by Jacob Christian Schaeffer.

The compiler of this bibliography has never seen but one copy of this edition and that in his own library. It is a crudely printed pamphlet comprising 32 pages of text and one hand-colored engraving of vegetation, including cat-tails, earthmoss, and "wilde wyngaard." There are 18 specimens of paper which embrace sheets made from nettles, straw, earth moss, cat-tails, and aloe leaves.

SCHAEFFER, JACOB CHRISTIAN, 1718-1790.

Erweis in Musterbögen dass die neuen Papierarten von Papelwolle, Baumblättern, Hanfagen, u. s. w. sich allerdings auch zu Tapeten übermahlen und gebrauchen lassen. Regensburg, 1772.

Issued by Dr. Schaeffer to show the use of his paper for tapestry painting.

DESMAREST, NICHOLAS, 1725-1815.
Mémoire sur la papeterie en Hollande. In: Mémoires de l'Académie des Sciences. Paris, 1771-74.

Dr. JACOB CHRISTIAAN SCHÄFFERS

PROEFNEMINGEN

EN

MONSTER-BLADEN,

OM

PAPIER TE MAAKEN

ZONDER LOMPEN,

OF

Met een gering Byvoegzel derzelven.

TWEEDE DEEL.

Uit het Hoogduitsch vertaald.

Met Achttien Papier-Monsters en Eene gekleurde Plaat.

Te AMSTERDAM,
By JAN CHRISTIAAN SEPP, Boekverkoper.
MDCCLXX.

The title-page from the Dutch edition of Schaeffer's work, published in Amsterdam in 1770. This volume is described on page 37 of this book.

VON MURR, CHRISTOPH GOTTLIEB, 1733-1811.
Journal zur Kunstgeschichte und zur allgemeinen Litteratur. Nürnberg, 1777-84. Volumes 5 and 13. 10 x 17 cm.

In volume 5 (pages 127-145) there is a description of paper and five early watermarks are reproduced, including an anchor from 1376 and a bull's head of 1388. In volume 13 (pages 48-53) a brief history of paper is given.

COMENIUS, JOHANN AMOS, 1592-1670.

Visible World: or, A Nomenclature, and Pictures, of all the Chief Things that are in the World, and of Men's Employments therein; In above 150 cuts. Written by the author in Latin and High Dutch . . . Translated into English by Charles Hoole, M. A. London, 1777. 11 x 18 cm.

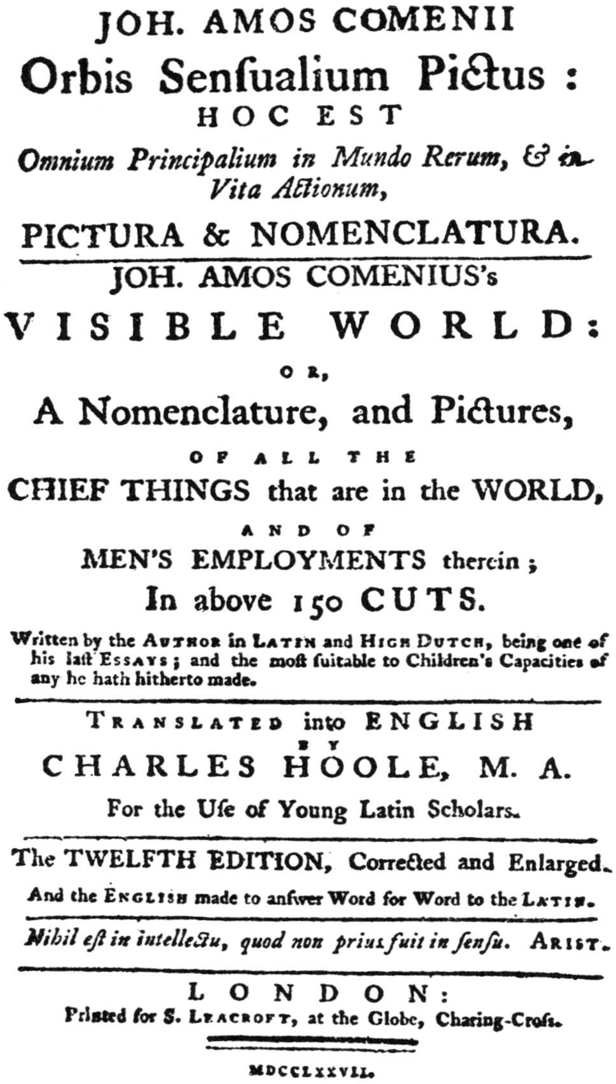

JOH. AMOS COMENII

Orbis Senſualium Pictus :

HOC EST

Omnium Principalium in Mundo Rerum, & in Vita Actionum,

PICTURA & NOMENCLATURA.

JOH. AMOS COMENIUS's

VISIBLE WORLD:

OR,

A Nomenclature, and Pictures,

OF ALL THE

CHIEF THINGS that are in the WORLD,

AND OF

MEN'S EMPLOYMENTS therein ;

In above 150 CUTS.

Written by the AUTHOR in LATIN and HIGH DUTCH, being one of his laſt ESSAYS ; and the moſt ſuitable to Children's Capacities of any he hath hitherto made.

TRANSLATED into ENGLISH BY CHARLES HOOLE, M. A.

For the Uſe of Young Latin Scholars.

The TWELFTH EDITION, Corrected and Enlarged.

And the ENGLISH made to anſwer Word for Word to the LATIN.

Nihil eſt in intellectu, quod non prius fuit in ſenſu. ARIST.

LONDON:

Printed for S. LEACROFT, at the Globe, Charing-Croſs.

MDCCLXXVII.

A facsimile of the title-page of the 1777 edition of Comenius' work.

The first edition of this volume was the earliest pictorial schoolbook issued and, like the numerous subsequent editions, attempted to explain the "chief things that are in the world, and of men's employments."

The text of the first edition was written in Latin and High Dutch, "every where one word answer'd to another over against it."

Papermaking is illustrated by the quaint wood-engraving here reproduced. The description refers to the numbers on the picture and reads: "The ancients used beech-boards, 1, or leaves, 2, also bark of trees, 3; especially of an Egyptian shrub which was called papyrus.

"Now paper is in use, which the papermaker maketh in a paper-mill, 4, of linen rags, 5, stamped to a mash, 6, which being taken up in frames, 7, he spreadeth into sheets, 8, and setteth them in the air so that they may be dried. Twenty-five of these make a quire, 9, twenty quires a ream, 10, and ten of these a bale of paper, 11. That which is to last long is written on parchment."

GUETTARD, JEAN ÉTIENNE, 1715-1786.

An enquiry concerning the materials that may be used in making paper. In: Select Essays: Collected from the Dictionary of Arts and Sciences, and from various modern Authors. Philadelphia, 1777. 11 x 19 cm.

This essay was the first article to be published in America on the subject of papermaking materials. This is the same treatise that was published in London in 1754, and described on page 31 of this bibliography.

SELECT ESSAYS:

CONTAINING:

The Manner of raising and dressing Flax, and Hemp.	The Manner of raising Radishes, Turnips, Cabbage, and other such Plants.
ALSO, The whole Method of Bleaching or Whitening Linen-Cloth.	AND an Enquiry, concerning the materials that may be used in making Paper.
LIKEWISE, Observations on the Management of Cows and Sheep.	WITH, Valuable Dissertations on other useful Subjects.

Collected from the Dictionary of Arts and Sciences, and from various modern Authors.

Felix, qui potuit rerum cognoscere causas.

Happy, he who can trace out the Causes of Things.

GEORGIC 2. DAVIDSON'S VIRGIL.

PHILADELPHIA:

PRINTED, BY ROBERT BELL, next Door St. PAUL'S CHURCH, in Third-street, Philadelphia.

M,DCC,LXXVII.

Title-page of the first American treatise on paper materials.

LALANDE, JOSEPH JÉRÔME LE FRANÇAIS DE.

Arte de hacer el papel según se practica en Francia, y Holanda, en la China, y en el Japón. Traduc. por. Mig. Geron. Suarez y Nuñez. Madrid, 1778.

The Spanish translation of the textbook by De La Lande, first published in French.

VON MURR, CHRISTOPH GOTTLIEB, 1733-1811.

Beschreibung der vornehmsten Merkwürdigkeiten in Nürnberg und auf der hohen Schule zu Altdorf. Nürnberg, 1778. 10 x 17 cm.

VON STETTEN, PAUL, 1731-1808.

Kunstgewerb und Handwerks Geschichte der Reichs-Stadt Augsburg. Augsburg, 1779. 10 x 17 cm.

DIDOT, FIRMIN, 1764-1836.

Lettre sur les découvertes de M. Didot, aîné, dans les arts de la papeterie. Paris, 1783.

A pamphlet of sixteen pages relating to the discoveries of M. Didot, the elder, in the art of paper manufacture.

BREITKOPF, JOHANN GOTTLOB IMMANUEL, 1719-1794.

Versuch, den Ursprung der Spielkarten, die Einführung des Leinenpapieres, und den Anfang der Holzschneidekunst in Europa zu erforschen. Leipzig, 1784-1801. 2 volumes. 19 x 24 cm.

This two volume work is printed on thin, unsized paper of a blue-grey color. Most of the contents is devoted to the history of playing-cards and engraving on wood. The first volume (pages 45-112) gives a description of linen paper with engravings showing the interior of a Japanese paper mill, and the appliances of a Hindustani paper manufactory. There are also engravings of pulp stampers and reproductions of twenty early watermarks, mostly bull head devices of the fourteenth and fifteenth centuries.

Breitkopf was a celebrated printer and publisher of Leipzig, and also a skilled musician. In 1750 he invented movable types for printing music.

ASTLE, THOMAS, 1735-1803.

The Origin and Progress of Writing . . . also some account of the Origin and Progress of Printing. London, 1784. 22 x 29.

Pages 203-207 are devoted to the history of papyrus, Chinese paper, cotton paper, and paper made from linen rags.

VILLETTE, CHARLES MICHEL DE, 1736-1793.

Oeuvres du Marquis de Villette. London, 1786. 10 x 16 cm.

The paper upon which this edition was printed was made from the bark of the lime tree; another edition on marsh mallow paper. The maker of the paper was Lévrier de Lisle, who in a prefatory note says he has used different plants, barks, and common vegetables for making paper, as he desired to prove that less valuable material could be substituted for that in general use and which was daily becoming rarer. At the end of the book there are specimen pages of paper made from nettles, hops, moss, reeds, three varieties of conferva, roots of dandelions, hazel wood, prick wood bark with its pellicle and crust, marsh mallow, poplar, oak, and osier bark.

SCHWANDNER, JOHANN GEORG.

Chartam lineam antiquissimam omnia hactenus producta specimina aetate sua superantem ex cimeliis bibliothecae Augustae Vindobonensis omnium Europae eruditorum judicio exponit Joannes Georgius Schwandnerus. Vindobonae, 1788. 18 x 24 cm.

This thin quarto volume describes a fragment of paper supposed to have been the earliest linen paper discovered to that time. The sheet was found among the charters of the ancient Monastery of Göss in Upper Stiria, and thought to have been from about 1243.

Schwandner was Principal Keeper of the Imperial Library, Vienna, where the book was printed.

DESMAREST, NICHOLAS, 1725-1815.

Art de fabriquer le papier. In: Des arts et métiers mécaniques de l'encyclopédie méthodique. Paris, 1788. Volume 5.

WEHRS, GEORG FRIEDRICH, 1750-1818.

Vom Papier, den vor der Erfindung desselben üblich gewesenen Schreibmassen und sonstigen Schreibmaterialien. Halle, 1789. 12 x 21 cm.

Vom

Papier,

den

vor der Erfindung desselben

üblich gewesenen

Schreibmassen,

und

sonstigen

Schreibmaterialien.

Von

Georg Friedrich Wehrs.

Halle,

bey Johann Jacob Gebauer,

1789.

A reproduction of the title-page of the work by Georg Friedrich Wehrs.

This compilation remains one of the most comprehensive volumes on the history of early writing materials, including papyrus, parchment, and paper, that has been written. During the year 1790 a supplement to this book was compiled by Wehrs and published in Hannover.

HOMER, ARTHUR, 1758-1806.

Bibliotheca Americana; or, A chronological catalogue of the most curious and interesting books, pamphlets, state papers, etc. upon the subject of North and South America, from the earliest period to the present.
London, 1789. 21 x 26 cm.

An interesting paragraph relating to papermaking appears in this volume which reads: "The people of North America manufacture their own paper, and in sufficient quantities for home consumption; but the price of labour is so high as to discourage publishing beyond their own laws, pamphlets, and newspapers."

MEMORIAL

RELATIVE TO THE

INVENTION

OF A

NEW METHOD OF BLEACHING.

SHOWING THE ABSURDITY OF ANY
PRETENSIONS TO AN EXCLUSIVE
PRIVILEGE FOR USING IT

IN THE

PAPER MANUFACTURE.

EDINBURGH:

PRINTED BY WILLIAM SMELLIE.

M,DCC,XCII.

A facsimile of the title-page of the rare pamphlet on bleaching.

BLEACHING OF PAPER.

Memorial relative to the invention of a new method of bleaching, showing the absurdity of any pretentions to an exclusive privilege for using it in the paper manufacture.
Edinburgh, 1792. 13 x 20 cm.

This rare pamphlet of 43 pages gives a short account of the bleaching of paper, the invention of Karl Wilhelm Scheele (1742-1786), a celebrated German chemist, who lived in Sweden. The booklet also points out the absurdity of one firm (Messrs. Taylors) having the control of the process in the British Isles.

LALANDE, JOSEPH JÉRÔME LE FRANÇAIS DE, 1732-1807.

Volledige beschrijving van alle konsten, ambachten, handwerken, fabrieken, trafieken, derzelver werkhuizen, gereedschappen, enz. ten deele overgenomen uit de beroemdste buitenlandsche werken... de papiermaaker.
Dordrecht, 1792. 13 x 23 cm.

The Dutch edition of this important eighteenth century textbook on papermaking. There are fourteen folded engravings showing the complete process of the craft. The engravings are identical with those of the French edition (1761) except for slight modifications in the costumes of the workers.

FRANKLIN, BENJAMIN, 1706-1790.

Description of the process to be observed in making large sheets of paper in the Chinese manner, with one smooth surface. In: Transactions of the American Philosophical Society . . . for promoting useful knowledge.
Volume 3, pages 8-10. Philadelphia, 1793. 19 x 25 cm.

This communication by Dr. Franklin was read before the Society June 20th 1788.

Dr. Franklin describes in his essay the laborious method used in Europe for forming large sheets of paper, and then he relates the simple manner employed by the Chinese papermakers in solving the difficulty.

The European method consisted of pasting small sheets together at the edges and burnishing the joints with a flint. In China a large mould was used, operated by two men and balanced by a counterpoise. The large sheets were couched upon the flat inclining sides of a heated kiln which gave one side of each sheet a remarkably smooth surface. Dr. Franklin states that the Chinese workmen could make paper "four and an half ells long by one and an half ells wide" by their method.

VOLLEDIGE
BESCHRIJVING
VAN ALLE
KONSTEN, AMBACHTEN, HANDWERKEN, FABRIEKEN, TRAFIEKEN, DERZELVER WERKHUIZEN, GEREEDSCHAPPEN, ENZ.
TEN DEELE OVERGENOMEN
UIT DE BEROEMDSTE BUITENLANDSCHE
WERKEN;
En vermeerderd met de Theorie en Praktijk der beste Inlandsche Konstenaaren en Handwerkslieden.
NEGENDE STUK.
DE
PAPIERMAAKER.
MET PLAATEN.
Te DORDRECHT,
BIJ A. BLUSSE EN ZOON.
MDCCXCII.

The title-page of the Dutch edition of the textbook by De La Lande.

SCHWARZ, CHRISTIAN GOTTLIEB.

Explication des filigranes qu'on trouve dans les papiers des premiers livres imprimés à Mayence. Tiré des Opuscula quaedam academica varii argumenti. Nürnberg, 1793.

DENNE, SAMUEL, 1730-1799.

Observations on Papers. In: Archaeologia. Volume 12. London, 1795.

Pages 114-131 are devoted to Rev. Samuel Denne's discourse on paper. There are four plates of various ancient watermarks and one plate of Pott marks, dating from 1604 to 1663.

An advantageous Hint :

TO THE

EAST INDIA COMPANY, CAPTAINS, OFFICERS, SUPERCARGOES, IMPORTERS OF RICE AND INDIA BALE GOODS. *Also* GROCERS, DRAPERS, GUNPOWDER MAKERS, *and* PAPER MAKERS *in general.*

THE PAPER ON WHICH THIS IS PRINTED is manufactured from an East India Article, called PAUT or JUTE (CROTALARIA JUNCEA, or PAUT) *, which grows in India; it is the same from which Gunney Bags, or such as bring over Sugar, Salt Petre, Pepper, Bale Goods, &c. are made, and may be collected from most Grocers, Drapers, and Gun Powder Makers, &c. &c. in England, at a small expence (*as old bags are better than new for this purpose*). The first idea was communicated to J. SEWELL, of Cornhill, by an ingenious Literary Gentleman, long resident in India, on account of an advertisement which appeared on the Covers of the EUROPEAN MAGAZINE (Addressed to Ladies, &c. not to destroy their Linen Rags), by J. SEWELL, No. 32, Cornhill; who takes this method of recommending to Paper-Makers in general the manufacturing a useful paper (demy, crown, or cartridge) for the use of Grocers, Chemists, &c. which will greatly decrease the consumption of rags, and of course the price of paper.

The plant which yields the fibre from whence this flax-like substance is obtained, is called, by the natives of Bengal, *Paat*, with additional names to distinguish the several varieties.—The *Fibre* is called *Jute*.—It is much cultivated in Bengal for the making of a coarse sacking (called Gunny), ropes, twine, &c. but is now used for marine purposes—nor is paper manufactured from it in Bengal.

The cost of this article, in Bengal, is about 5s. per cwt. and it has sold in England at 20s. and 29s. per cwt.

The two following varieties have been cultivated in the Botanical Garden at Calcutta, viz.

Bhungee Paat.—Corchorus Olitorius ; Linn. and,
Ghee Naltha Paat.—Corchorus Capsularis, Linn. †

HEMP.

The substitute for Hemp in Bengal is the fibre of a plant, called *Sun* (Crotolaria Juncea, Linn.), by which name the fibre is also known

* There is a small quantity of bleached coloured rags mixed with it; or if wanted of an inferior colour as wrappers for variety of goods, cartarage, literas, or even blue, it may be made of the Gunney Bags only, and will then rank in the lower duty as a wrapping paper. *Samples of Paper made from Gunney Bags only may be had at the Neckinger Mill, Bermondsey, or at their Warehouse, No. 376, Strand.*

The first page of the pamphlet regarding jute for papermaking, described on page forty-six.

LEMOINE, HENRY, 1756-1812.

Typographical Antiquities. History, Origin, and Progress of the Art of Printing ... with a Curious Dissertation on the Origin and Use of Paper. London, 1797. 10 x 17 cm.

JUTE FOR PAPERMAKING.

An advantageous hint: To the East India Company, Captains, Officers, Supercargoes, Importers of Rice and India Bale Goods. Also Grocers, Drapers, Gunpowder Makers, and Paper Makers in general.
London, 1797. 14 x 22 cm.

This pamphlet of sixteen pages consists mainly of advertisements for the publications of J. Sewell, 32 Cornhill, London. The first two pages are devoted to a communication by Mr. Sewell relative to the use of jute as a substitute for rags in papermaking. This rare pamphlet is printed upon paper made from jute or paut found in East India.

A facsimile of the first page of the pamphlet is given on page 45 of this bibliography.

CAMUS, ARMAND GASTON, 1740-1804.

Notice d'un livre imprimé a Bamberg en mccccclxii, lue a l'Institut National.
Paris, 1799. 30 x 23 cm.

Regarding a book imprinted at Bamberg in 1462. Reproductions of three fifteenth century bull's head watermarks are given.

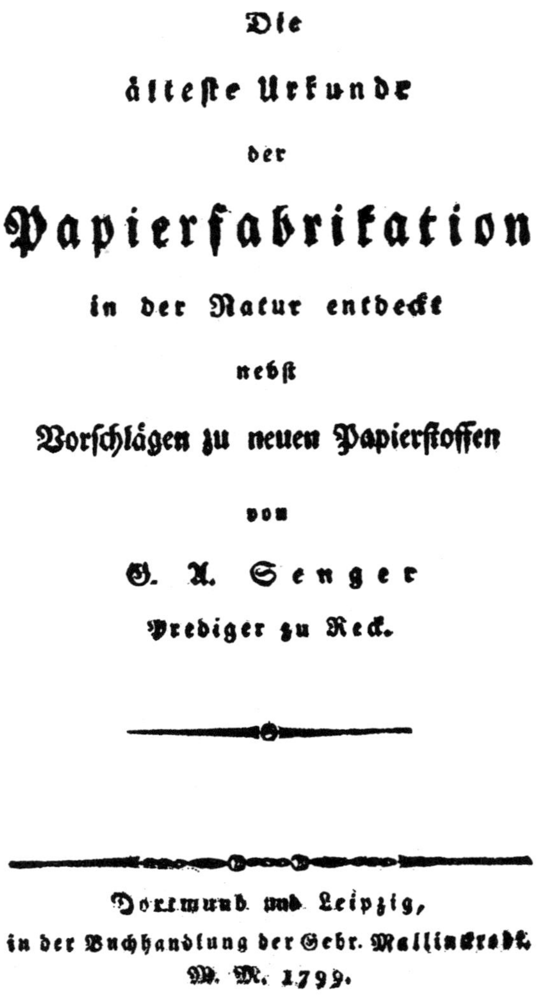
Die
älteste Urkunde
der
Papierfabrikation
in der Natur entdeckt
nebst
Vorschlägen zu neuen Papierstoffen
von
G. A. Senger
Prediger zu Reck.
Dortmund und Leipzig,
in der Buchhandlung der Gebr. Mallinckrodt.
M. M. 1799.

A facsimile of the title-page of G. A. Senger's rare work on papermaking in nature.

SENGER, G. A.

Die älteste Urkunde der Papierfabrikation in der Natur entdeckt nebst Vorschlägen zu neuen Papierstoffen von G. A. Senger, Prediger zu Reck.
Dortmund and Leipzig, 1799. 13 x 19 cm.

An exceedingly rare book of ninety-six pages relating to the oldest record of papermaking discovered in nature, with proposals for new materials for making paper. The essay is printed on paper fabricated from conferva, a water plant, called by Senger water wool. He states that water wool, or river paper, was the oldest form of papermaking in nature. This material he termed a coralline product, being the web of water insects, seed capsules, or insects' eggs. Senger discusses the probability of sufficient conferva being found as a substitute for rags in papermaking.

HISTORICAL ACCOUNT

OF THE

SUBSTANCES

WHICH HAVE BEEN USED TO

DESCRIBE EVENTS, AND TO CONVEY IDEAS,

FROM THE

EARLIEST DATE

TO THE

INVENTION OF PAPER.

SECOND EDITION.

PRINTED ON PAPER RE-MADE FROM OLD PRINTED AND WRITTEN PAPER.

By MATTHIAS KOOPS, Esq.

LONDON:
PRINTED BY JAQUES AND CO. 30, LOWER SLOANE STREET.
1801.

The title-page of the second edition of the work by Matthias Koops, published in the year 1801.

KOOPS,
MATTHIAS.

Historical account of the substances which have been used to describe events, and to convey ideas, from the earliest date, to the invention of paper.
London, 1800.
16 x 27 cm.

The earliest edition of this notable work was "printed on the first useful paper manufactured solely from straw." The paper is of the laid type and bears the crown and G.R. watermark. There is an appendix of seven pages "printed on paper made from wood alone, the produce of this country, without any intermixture of rags, waste paper, bark, straw, or any other vegetable substance, from which paper might be, or has hitherto been manufactured." The first edition embraces eighty-two pages of text.

The second edition of Matthias Koops' work (London, 1801. 12 x 20 cm.) was "printed on paper re-made from old printed and written paper," with a sixteen page appendix on paper made from wood. The paper used in this edition is of the wove type, without watermarks. Another printing of this edition, the rarest of the three, was issued in 1801, and unlike the regular second edition, was printed on paper made from straw. Only a limited number of copies were so published.

An engraving (27 x 36 cm.) entitled: "Noon" by Austin, after J. Ruysdael appeared at the beginning of the nineteenth century with the caption: "On paper made from straw alone by Mr. Koops, 1800."

Although the books by Koops are the most modern volumes to be enumerated in this bibliography, the life and character of Matthias Koops remains the most obscure of any of the scientists listed. A thorough search in the American genealogical libraries, and also in the British Museum and Patent Office Libraries, revealed nothing concerning him. It has not even been possible to ascertain the date of his birth or death.

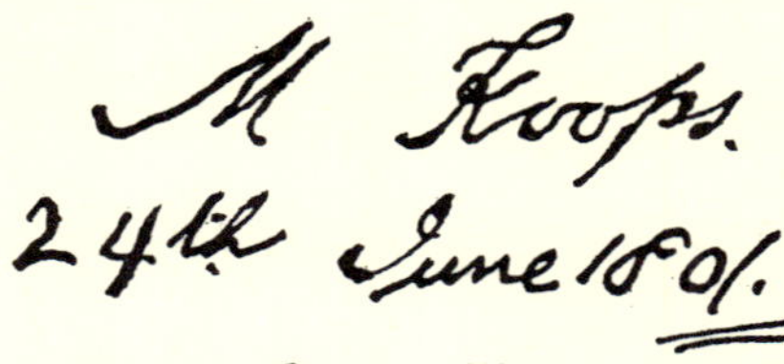

Matthias Koops' signature, 1801.

Koops' first writing was undertaken in 1796 when Jacques and Thomas of Chelsea, London, published for him an octavo volume entitled: "A development of the views and designs of the French nation, and the advantages which will derive to them, if they should be able to secure to themselves... the free navigation of the rivers Rhine, Maese, and Scheldt." This book was dedicated to the Duke of York, while the three editions of the work regarding paper were dedicated to George the Third.

Mr. Koops was the first man in England to be successful with material other than rags for papermaking. His experiments were accomplished at Neckinger mill, Bermondsey and the three following patents pertaining to his researches are recorded in the British Great Seal Patent Office:

"A. D. 1800, April 28th – No. 2392. Matthias Koops of Queen Street, Ranelagh in the county of Middlesex, gentleman, for a mode of extracting printing and writing ink from printed and written paper, and converting the paper from which the ink is extracted into pulp, and making thereof paper fit for writing, printing, and other purposes."

"A. D. 1800, August 2nd – No. 2433. Matthias Koops of Queen Street, Ranelagh in the county of Middlesex, gentleman, for a method of manufacturing paper from straw, hay, thistles, waste, and refuse of hemp and flax, and different kinds of wood and bark."

"A. D. 1801, February 17th – No. 2481. Matthias Koops of James Street, Westminster, gentleman, for a method of manufacturing paper from straw, hay, thistles, waste, and refuse of hemp and flax, and different kinds of wood and bark, fit for printing and other useful purposes."

The text of the three editions of Koops' work are practically identical and treat chiefly of the various materials that have been used in past ages as substances upon which to write.